CAREERS in CONSTRUCTION

PEARSON
Prentice
Hall

nccer

Upper Saddle River, New Jersey
Columbus, Ohio

National Center for Construction Education and Research

President: Don Whyte
Director of Revision and Development: Daniele Stacey
Production Manager: Jessica Martin
Product Maintenance Supervisor: Debie Ness
Editor: Bethany Harvey

Writing and development services provided by MetaMedia Training International Inc., Germantown, Maryland.

10 9 8 7 6 5 4 3 2
ISBN 0-13-228605-X

Dear Reader:

Congratulations. You've taken the first step in exploring what could become a very satisfying and rewarding career in the construction industry. I took that step myself years ago, and I'm here to tell you, getting involved in construction was one of the better decisions I ever made.

I actually started working construction at age 16, working summers for my dad, the owner of a small industrial construction company in Grand Junction, Colorado. I decided to pursue a carpentry apprenticeship, rather than attending college. After four years of training, I became a journeyman carpenter, earning all the while I was learning.

Within a few years, I worked my way from journeyman carpenter to foreman, general foreman, and then to superintendent. Eventually I started my own business: TIC—The Industrial Company. When I started the business in 1974, we had 17 employees and most of our business came from the construction of condominiums to water pipelines. From there we branched out into mining—mining everything from coal to precious metals such as gold. No project was too big for our sights.

Fast forward to today: TIC is a $1.2 billion dollar company with approximately 9,000 employees. How did we get here? Our success lies in the people, the employees of TIC. People in construction are among the most talented and visionary people there are.

The construction industry holds an abundance of career opportunities. Millions of people like me build rewarding careers and livelihoods in this industry each year. So, take a journey through this book and see what might lie ahead for you. And remember, no dream is too big, and no beginning too small.

Yours truly,

R.W. McKenzie

R.W. McKenzie

CEO and COB

TIC Holdings, Inc.

Acknowledgments

NCCER gratefully acknowledges the organizations and individuals that have contributed to the development of Careers in Construction.

Through our Success Stories features, we are pleased to showcase the achievements of Steve Bugden, Eric Carley, Shaun Fitzgerald, Kyle Fitzsimmons, Jared Garber, Elizabeth Kelly, Bill Kowalik, James Sewall, and Jennifer Smith.

Scott Shelar and the Construction Education Foundation of Georgia (www.cefga.org) provided individual success stories about successful competition winners Fitzgerald, Kelly, Kowalik, and Smith. Todd Staub and Associated Builders and Contractors (www.abc.org) provided success story candidates and national competition award winners Carley and Fitzsimmons, and ABC contributed many images to accompany the text. Associated General Contractors of America (www.agc.org) and SkillsUSA (www.skillsusa.org) provided advice and insight to our work.

Cal Pygott of the Construction Academy at Bothell High School in the Northshore School District outside Seattle, Washington, and Linda Lynch, Paul Jeffries, and Steve Boden of the Thomas Edison High School of Technology (TEHST), a part of the Montgomery County, Maryland, public school system, illustrated the education pathways that prepare students to enter the construction professions upon graduation. NCCER's training curriculum is used in both of these high schools. TEHST generously provided a number of images for our book.

Mark Prenni of Black and Veatch (www.bv.com) was our subject matter expert on safety and personal protection. Mark patiently and diligently reviewed every page and image through several revisions of the layout.

Careers in Construction
Table of Contents

Construction is everywhere you go—your home and school, where you, play, and shop, and the roads that get you there and back.

Construction is our biggest industry. It's a powerful engine that drives the American economy, employing more than 7 million workers and providing more than $500 billion of our annual domestic product.

The next time you drive around, notice all the structures being built—a new home or condo for a young family, a gleaming, 20-story office building, the remodeling of an older home, or a highway overpass to relieve traffic. Hard hats everywhere you look!

Construction affects everyone, everywhere. Without it, cities would stagnate; living conditions would deteriorate; and local, state, and national economies would begin to fail.

BUILDING A WORLD

What do the Parthenon, Eiffel Tower, Empire State Building, and White House have in common? They were all built by skilled construction workers.

From the time of the pyramids—from great cathedrals to great cities—construction workers have built every single structure that rises from the face of the earth. In each city's distinctive skyline you see the collective handiwork of thousands of construction workers applying their muscles and talent over decades, and even centuries.

HIGH DEMAND FOR CONSTRUCTION WORKERS

But with all the construction going on, there's a major problem: not enough workers. According to the Bureau of Labor Statistics, we need 250,000 new construction workers this year and each year after that to meet the demand for new homes, schools, office buildings, airports, industrial plants, and malls.

Construction: America's Powerhouse Industry

Construction workers are in high demand because they are skilled workers. They know what to do and how to do it. Under that hard hat is a mind that's been technically trained with the equivalent of a college degree—a valuable worker who can get a job just about anywhere in the country.

JOBS THAT WON'T GO AWAY

These days, construction workers feel secure because they know the demand for their jobs will remain strong for a long time. And they know that construction jobs aren't the ones being outsourced to other countries. "Hey," one Atlanta plumber says, "people are coming here for work—and there's plenty to go around."

So right now is a great time to think about a career in construction.

ABOUT THIS BOOK

Careers in Construction introduces the world of construction and the rich career opportunities it offers. We'll review the economic and other benefits of a construction job, and then look at the factors that would make you a good candidate for construction work.

Next, we'll take an extended tour of several work sites. You'll learn what the individual construction trades do, and how important teamwork is to the successful completion of each project. Along the way, you'll meet construction workers—from recent high school grads to company owners—who will tell you why construction is good for them.

Finally, we'll explore the various construction career paths that are open and waiting for you right now. We'll show you how to get started and where to look for educational and job opportunities. There's never been a better time to proudly wear a hard hat, so let's get started!

Success STORIES

THE CARPENTER WHO "WOOD" BE HAN SOLO

Indiana Jones was first a carpenter?

That's right, **Harrison Ford** of *Star Wars* and *Indiana Jones* is a self-taught master carpenter—a hobby he still enjoys today.

While struggling to become a working actor in Los Angeles, supporting his wife and two small sons, Ford learned the carpentry trade from books. He had a reputation as one of the best cabinetmakers in Los Angeles, and his services were in high demand on the trendy west side long before he became a movie star.

Coincidentally, it was Ford's carpentry that led him to his breakout role of Han Solo in *Star Wars* in 1975. While casting the movie, Director George Lucas hired Ford to build cabinets in his home. Lucas asked Ford to read lines for actors auditioning for the roles in *Star Wars*. The rest is history.

Today Ford lives in a white-painted ranch house that he built himself in Jackson Hole, Wyoming.

DOES OZZY HAVE THE PIPES?

What does rock star Ozzy Osbourne have in common with mathematician genius Albert Einstein?

One worked as a plumber and the other wished he had. Can you guess which? Osbourne left behind his plumber's wrench to be the front man for heavy metal band Black Sabbath. It was Einstein—the level-headed one—who remarked wistfully that if he had to do it all over again, he 'would choose to be a plumber.' Smart man.

MAKING THE MOST OF AN OPPORTUNITY

You could probably call James Sewall a bit of an opportunist. A soldier turned craftsman, Sewall has always looked for the opportunity to do more, be more, and achieve more. "I've always thrown my name in the hat for extra work," he said. "I'd take any opportunity I had to get exposure and build experience. And it makes you really stand out to an employer as someone who is on his way up."

The 40-year-old has been in the construction trades for more than 15 years, following his discharge from the U.S. Marine Corps and tour of duty in the Persian Gulf War. He began as a helper learning whatever he could on the job, became a boilermaker, and today is a project manager for Austin Industrial in Texas. "Back in those days, there really wasn't the education and training available that there is today," he said. "It was really all about on-the-job training."

Sewall is a self-proclaimed go-getter, never wanting "to be second fiddle to anyone."

"If you have the desire to work, the desire to excel, you'll succeed," said Sewall. So whenever there was an opportunity to work extra time and get experience on other projects, he took it. In 2003, he was awarded the Associated Builders and

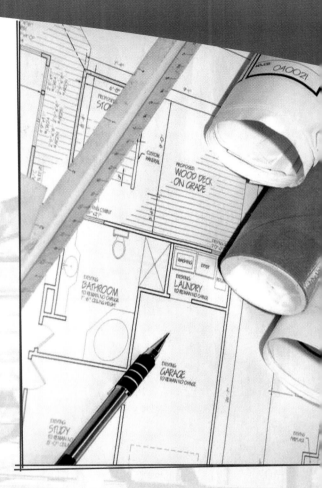

Contractors' Craft Professional of the Year for not only mastering his skills and taking pride in his work, but also for the example he sets for the next generation of craft professionals.

Sewall has previously been a core curriculum instructor for the National Center for Construction Education and Research (NCCER) and an instructor in the course for boilermaker assessment.

At the time Sewall was named Craft Professional of the Year, he was working as a boilermaker and set a goal at that time to become a project manager within 3 years. He achieved that goal recently and has managed his first project. He added that he enjoys managing as much as he did being a hands-on craftsman. Although he still needs the physical stamina to do the job, now he also gets to work more at mentoring and team-building.

Working to dispel the myth that money can't be made in the construction industry, Sewall noted that even before becoming a project manager, he was earning a six-figure income even without a college degree. But the training—and the learning—hasn't stopped, he said. "I'm still in training."

Although he has spent a great deal of time moving up in his career and taking on extra projects, Sewall has managed to maintain balance in his life. He raises horses on a small ranch he owns in Conroe, Texas, where he and his wife raise a 4-year-old daughter. They take time to travel the world and scuba dive together, he said. And he doesn't have a hard time leaving work at work.

"Once I leave these gates, I leave it all behind," he said.

So what's Sewall's 3-year goal today? "I'd like to be doing what I'm doing only on bigger projects," he said. "There's a lot more people to manage and much more challenge to bigger sites. It's a challenge to manage all of the different personalities."

> **IF YOU HAVE THE DESIRE TO WORK, THE DESIRE TO EXCEL, YOU'LL SUCCEED.**

First, let's put a myth to rest—the one that says you've got to have a college degree to make it in the world. It's not true, especially when you read that many 4-year college grads are having trouble finding jobs (other than waiting tables while they look for the "right" position).

TTO COLLEGE OR NOT TO COLLEGE? The reality is that a high school graduate with some technical training—perhaps through an apprentice program or a community college—may have a better chance of landing a good-paying job than a person with a bachelor's degree.

And here's another myth to knock down: that the only people entering the construction trades are those who can't or are unwilling to go to college. Definitely <u>not true</u>. Today, many students going to community college or a technical school already have a 4-year degree—they've seen the light and one of the brightest is a career in construction.

Researchers are finding that it is a worker's skills, rather than what type of diploma they have, that is the key to earnings differences. Earnings for technically skilled workers—such as journeymen electricians and heavy equipment operators—often exceed those of college grads with degrees in "soft" skills such as the humanities or social sciences.

More than ever, students, counselors, and parents are beginning to see construction as an attractive alternative for gaining the advanced skills and education that lead to rewarding, high-paying careers.

Consider a Career in Construction

While it's true that college grads make more on average than other workers, there is a wide range in what both groups get paid. A recent survey showed that the highest paid high school-diploma workers—many of them construction workers—were making twice the salary of lower-paid college grads.

A construction industry career makes good economic sense. It offers great pay, excellent benefits, and jobs that are always in high demand. A Bureau of Labor statistic showed that a skilled journey person earned nearly 174% more than a high school dropout, 100% more than a high school graduate, and 59% more than an associate degree holder.

Something else to consider: construction workers earn while they learn. During an on-the-job apprenticeship, a person may earn upwards of $80,000 over four years, while the college student is spending that much or more to earn a degree, and probably rolling up years of student loan payments.

SALARIES OF IN-DEMAND CONSTRUCTION OCCUPATIONS
(Source: Bureau of Labor Statistics)

Occupation Title	Hourly Wage Range	
	Median	Top 10%
Carpenter	$17	$29
Electrician	$20	$34
Brick and block mason	$20	$30
Plumber, pipefitter	$20	$34
Sheet metal worker	$17	$30
Heating, Venting, AC, and Refrigeration worker	$17	$27
Drywall installer	$16	$28
Crane and tower operator	$18	$28
Painter	$11	$18
Heavy equipment operator	$18	$28
Glazier	$16	$30
Millwright	$21	$32
Roofer	$15	$26
Welder	$15	$22
Structural steel worker	$20	$34
Construction laborer	$12	$24
Cost estimator	$24	$41
Supervisor	$18	$29
Construction manager	$34	$61

M

MOVING UP THE LADDER. Construction offers many opportunities and paths for gaining the additional skills and experience you need to advance to more responsible—and higher paying—positions.

Those entering the construction industry out of high school usually start as laborers, helpers, or apprentices. While some laborers and helpers can learn their job quickly, the skills needed to become a trades professional take years to learn through classes at a technical or trade school, an apprenticeship, or some other employer-provided program.

Apprenticeships usually last from 3 to 5 years and consist of on-the-job training and formal classroom instruction. Some apprenticeship programs now accept competency in place of time requirements, so you can complete a program in less time.

Construction trades workers of all types also gain additional experience and know-how by working on different types of projects such as housing developments, office buildings, and highway construction.

Flexibility and the willingness to adopt new techniques—plus the ability to be a team player—are essential for advancement. Those who have proven skills in many areas and show leadership qualities may become supervisors and managers of even larger projects. They may also transfer to jobs as building inspectors, sales representatives, or technical school instructors.

College grads entering construction usually start as management trainees. Others may begin as field engineers, schedulers, or cost estimators. With hard work, a team approach, and increasing experience, they may advance to positions such as construction manager, top executive, or independent contractor.

In most industries, how far you advance is entirely in your hands. This is especially true in construction, where everyone can see what you can do with your hands.

SKILLS THAT LAST A LIFETIME

Learn to ride a bike as a kid and you'll never forget the skill. It's the same in construction—learn to do a professional job, contribute to a team, and envision a completed project and the ability will never leave you.

These skills are like money in the bank. They can be taken out anytime, anyplace to support you and your family, and to add pleasure and a sense of accomplishment to projects you do for yourself and others.

Consider a Career in Construction

WHAT ARE WE DOING TODAY?

This is a common question in construction. Unlike many jobs where you do the same thing day in and day out, construction work is different—it brings new challenges every day.

When it rains, when the supplier is late, when a worker is absent, when the client wants a change—all these circumstances may affect your job. You may be asked to help out in a different job or even at another job site. Look on these challenges as opportunities to learn more and show your value as a team player.

CONSTRUCTION WORK MEANS TEAMWORK

We've mentioned the "team" concept several times so far, and there's a good reason. More than in most industries, construction depends on the teamwork of its workers to achieve success.

Houses, office buildings, and highway projects require people of many different talents—laborers and helpers, skilled trades professionals, truck drivers, engineers, and schedulers. For success they must all blend their skills into an efficient team to get the job done.

Think of a building project as a play. Each act advances the story and sets the stage for the next action. The actors— each with a unique role—skillfully tell the story and build toward the play's climax. Off-stage, they are supported by stagehands, set designers, the director, and dozens of other talents. And then the play ends. If all went right—if every team member did his or her job professionally—there is thunderous applause. It's the same with a building.

I **"I BUILT THAT!"** Most workers are proud of their accomplishments, especially when they've worked hard to gain their skills. Construction professionals are the same, but they have an advantage in the pride department.

While others may recall a big sale or other job success, construction people can point to their handiwork—perhaps an elegant staircase or award-winning brick design—and say, "I built that!" And, hopefully, it will still be there for the grandkids to admire.

In construction, your work, like Mr. Eiffel's, becomes a part of the fabric of the community and a lasting legacy to your skills and talent. As they say, you will be known by your works.

HAVE HARD HAT, WILL TRAVEL

Construction is going on all over the world, but only a few construction companies have the means and skilled workers to take on big projects in other countries. If your career path leads you to one of these industry giants, you may get the chance to work overseas, meet new people, learn about new cultures, and help make the world a better place.

OPEN FOR BUSINESS

Have you ever noticed how many commercial trucks and vans carry people's names —Turner Electrical Co., Johnson Plumbing, and Hernández Heating and Air Conditioning? You can be sure these successful entrepreneurs each started at the bottom. But through hard work, constant learning, and taking professional pride in their work, they were able to realize their dreams.

The opportunity to go into business for yourself is better in construction than most other industries. As a construction professional, you need only a moderate bankroll to become a contractor. You can work out of your home and hire other skilled craftspeople as needed for specific projects. You must be aware, however, that the contract construction field is highly competitive—taking a few business courses will increase your chance of success.

ALL IN THE FAMILY

Nearly 35 years ago when 18-year-old Steve Bugden told his father he wanted to be an electrician, his skeptical dentist-father replied, "What are you going to do, fix TVs?"

Bugden, now the owner of Steve Bugden Electric in Gaithersburg, Maryland, has long since dispelled the stigma that electricians have few options available to them and don't have much earning power. After spending time as an electrician's helper and apprentice, and then taking a short detour and attending community college to become a police officer, Bugden quickly decided that he'd rather stick to electrical work. While working as an apprentice for a local electrical company, Bugden took classes, and became licensed as a journeyman electrician within 4 years. By the time he was 26, he began working only 4 days a week for his employer while freelancing on his own one day a week. And as time went on, his customer base grew.

"I was lucky that I had a really flexible employer that allowed me to do that," Bugden said. "If I didn't have something set up for myself, I was able to do something for my employer. Suddenly, I found that it was more lucrative working for myself a few days a week than it was working for someone else all week."

Although Bugden admits he started his own business "by the seat of my pants," he credits his wife, Judy, as one of his biggest supporters. "She inspired and encouraged me from the start to go out on my own. She set up the bookkeeping and answered the phone while I worked out in the field. I couldn't have done it on my own. We started this business together and any success is certainly due to her help and support."

Bugden said he took the best feature from every company he has ever worked for and incorporated it into his own: show up on time, be honest, be organized and neat on the job, and be customer-friendly.

So as time went on, Bugden was able to hire additional electricians to go out in the field and he managed his business from the office. "Sadly, the construction field has not always attracted top-notch people," Bugden said. He hasn't always been pleased with his technicians' quality of work or their work ethic and habits. So since about 2000, Bugden scaled down the size of his business so that he works in the field and employs only a helper. "People who excel tend to want to get their own thing going," he said. He often loses skilled workers when they open their own businesses.

Although Bugden's business has focused primarily on retrofitting electrical work into existing homes and upgrading systems, he noted the many areas electricians can focus on. "Everybody needs to find their niche," he said, such as working on big commercial buildings, shopping malls, and office buildings; or specializing in high-rise office buildings for people who prefer not to work directly with customers.

"It's really changed from 35 years ago," Bugden said. "Technology has changed everything from security systems to audio–visual equipment; it's so technical you need someone with marketable skills to do it. There's very good money to be made."

Bugden's efforts may get passed on to his son, who has a degree in mechanical engineering. Working as a design engineer, Adam has found himself working mostly behind a computer. "But he's a more hands-on type of guy and a good problem solver," said Bugden. "So he has the hope and dream of taking over the business one day. And he can take it wherever he wants."

If you're a gifted athlete, you may seek a career in sports. Good at talking to people? Sales or management might be the thing. Want to face ultimate challenges? Try the military.

A construction career offers all of the above. Really. As a construction professional, you're part athlete as you test the strength and agility of your body dozens of times a day. You're the cool head stepping in to calm a work-site dispute. You're the project manager as you determine how best to deploy your "team" and assign tasks to complete a project on time and on budget. And you're a civilian warrior as you face the risk of handling high-voltage lines.

So how about you? Do you have what it takes to succeed in construction? Let's see.

Is Construction for You?

DO YOU ENJOY:

Hands-on work? Construction is the most hands-on profession in the world. On the job, your hands are the ultimate tool as you bend, cut, form, weld, nail, fasten, finish, and connect materials to build a structure that may last for centuries. As an advertising slogan might put it, "If you like working with your hands, you'll love construction." Some construction trades can even be considered art in the form of award-winning buildings.

Variety in what you do? "Hey, Dolores, can you help over here this morning?" As a construction worker, get used to the unexpected. Remember that construction is a team effort, and you may be asked at any time to lend a hand where it's needed to get the job done— look at it as another opportunity to show your ability and learn new skills. Construction and variety go together like a nut and a bolt.

Working with tools? Most of us have used a hammer, screwdriver, saw, or pliers at one time or another. Would you like a job where you used dozens of specialized hand tools and high-tech power tools every day? Construction might be for you. The work is challenging, and the industry is quick to put the best and latest tools into the hands of its craftspeople so they can do their jobs better and more efficiently. In some of the trades, you can expect to learn a lot about computer technology and computerized machinery.

Physical activity on the job? Construction is the exact opposite of a desk job. Skilled craftspeople are on their feet a lot, performing their specialized tasks, checking in to see what's next, taking an active part in making the architect's vision become a reality. It's not listed as a job benefit, but the workout you'll get during a construction day is like a session at the gym.

W

Working without supervision? Helper, apprentice, journey craftsperson, and on up the ladder—the farther you advance in the construction trades, the more you'll work on your own. Your coworkers and supervisors will come to trust your ability to do a professional job, until the only person you have to satisfy is yourself. Is this you? Eventually you may find yourself in a construction management position as a supervisor or foreman. You'll be able to pass on the knowledge you have gained working independently to younger coworkers.

Working outdoors? Let's face it—construction is an outdoor activity, especially at the start of a project. Depending on the location, you'll get hot, cold, and rained on as part of the job. But are you the kind that gets a gritty satisfaction from doing a professional job no matter what the obstacles? Welcome to construction.

Working Indoors? Construction can also be an indoor activity. Providing the electrical conduit to power up a skyscraper, reinforcing the walls, installing sprinkler systems and other safety systems, putting up the satellite receivers, and installing the heating and air conditioning systems—this is also construction.

Taking pride in your work? Are you the kind of person who stands back from a completed job and thinks, "Hey, I did that—good job"? If so, you should take a closer look at a construction career, where you'll get that chance several times a day. In fact, each thing you build becomes a mini-monument that can be admired for years to come. Take a good look at the St. Louis Arch or the Seattle Space Needle or the Statue of Liberty.

Is Construction for You?

WORK HABITS

When a construction company hires an employee, along with the person's skills they're also getting his or her work habits. Suppose you are that person. What kind of personal habits and work ethic will you bring to the job? <u>Are you:</u>

 Dependable? Employers, supervisors, and other workers like dependable people who come to work on time, have few absences, and take their responsibilities seriously. These people do their job correctly and finish it on time. Sound like you?

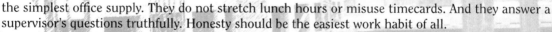

 Honest? There can be no question here—you're honest or not. Honest construction workers do not take others' tools or anything belonging to the company—even the simplest office supply. They do not stretch lunch hours or misuse timecards. And they answer a supervisor's questions truthfully. Honesty should be the easiest work habit of all.

 A team player? We keep stressing "teamwork" because it's the key factor in any successful construction project. It takes a skilled team of professionals—each member doing their part and helping one another—to turn an open space of ground into a modern, efficient building. Construction crews all over the country are looking for men and women who list "team player" among their professional skills.

 One who shows initiative? You want to give a fair day's work for a fair day's pay, but there is more you can do. If you can solve a problem yourself, do it—don't wait for someone else to tell you what to do. When you finish your job ahead of time, look for something else to do. And if you find a better, faster, or safer way to do a job, tell your supervisor. You'll be recognized as a person with initiative and an even more valuable employee.

 Willing to learn? The construction industry is changing every day with new techniques, tools, materials, and safety rules. Employers look for workers who not only learn everything they can about their jobs, but also absorb new information so they can do them more professionally. In fact, willingness to learn is a valuable trait no matter what your job is. Research shows that lifelong learners are happier and remain mentally sharp longer.

Success STORIES

HANDS-ON AT WORK

Eric Carley knows there will always be a need for people to build buildings, lay plumbing, service furnaces and air conditioners, and wire the electricity, "and that's a good feeling to know you'll always have a job," said Carley, who first got his start on a small town lumberyard, became a carpenter, and today is an assistant foreman with Silloway Builders, Inc., in Berlin, Wisconsin. "With all the tech classes I took (in high school) I had a good feeling that one day I would end up in the trades."

But when he started as a 19-year-old, he was a self-proclaimed party-goer, often staying up late and waking up early for work—and his work suffered. "Looking back I see that it was affecting everyone I worked with," he said. He made a turnaround when he enrolled in the Associated Builders and Contractors (ABC) apprenticeship program. "I took an active role to better myself and prove that I belong in the company. You learn a lot of the things you need on a daily basis

in the field, which, in turn, makes you more productive and gains you respect among your fellow workers."

Carley's efforts have paid off since he was named the state's top carpenter at the ABC Wisconsin State Construction Skills Competition. That award qualified him for the ABC National Craft Championship in Las Vegas, where he placed third. "It's a good feeling knowing only two apprentice carpenters in the nation are better than me."

Carley's competitive experience has mirrored that of fellow Wisconsin HVAC technician Kyle "Fitz" Fitzsimmons, who also placed first in the statewide competition and third at nationals.

At the national competition held in Las Vegas, Fitzsimmons was required to take a 2-hour paper-based test as well as an in-depth 6-hour practical test in which he had to work on a rooftop HVAC unit and troubleshoot problems with the unit and service it—all with a judge looking over his shoulder. "He'll come up and ask you questions and have you explain what you would do, why we do that, and if you can't find the information at the moment, where you would go to find it," Fitzsimmons said. "So basically, you're doing it hands-on and explaining at the same time. They wanted to make sure you actually knew what you were doing. So I found out how much I really knew about it, which is kind of interesting and I was surprised at how well I did overall."

But it's the hands-on aspect of his job that led Fitzsimmons to the construction trades.

"IT'S A GOOD FEELING KNOWING ONLY TWO APPRENTICE CARPENTERS IN THE NATION ARE BETTER THAN ME."

"I'm the kind of person who likes hands-on work," he said. "I'd rather not be in a classroom learning. I thought it would be a good career move for me."

When Fitzsimmons was in high school, he excelled as an athlete, truly enjoying the contact aspect of wrestling, football, and baseball. Even now he helps his parents run their 120-acre dairy farm in Middlepoint, Wisconsin, managing the herd and completing chores. So it only made sense that when a school speaker came to his high school to promote a career in the construction trades, he discovered his future path.

Fitzsimmons chose a career in HVAC because of all the options he had, and he has gone nowhere but up. After graduating high school and becoming an HVAC apprentice, he started at Madison Area Technical College, where he is now in his final year.

Most of the work Fitzsimmons and his shop perform is in the area of service—performing repairs on heating and air conditioning systems, mostly at residences. He makes a point of keeping up on new technology in the industry. "It's job security," he said. "You have to keep up on it to

keep going. The guys who go by instinct are getting phased out because they can't do the newer stuff. So that's where the newer kids are taking over who have gone through school have a big plus. Once you have your license you are guaranteed a spot almost anywhere you go if you have good work skills and work ethics."

BIG DREAMS AND GIANT LIGHT POLES

The giant light poles lay all around the Floyd College parking lot, under construction in Cartersville, GA. They are waiting for someone to stand them up and do the electrical wiring.

A shiny red pickup truck roars up to the curb. A 6'-3" giant of a man gets out. Yep. This is our guy. The light poles are going up. Bill Kowalik is on the job. Kowalik is only 19 years old. One year out of McEachern High School in Cobb County, he's already a fourth of the way to becoming a licensed electrician. "My company, Wayne J. Griffin Electric, has its own apprenticeship program. After four years of school and on-the-job training, I'll have my electrician's license," said Kowalik. "I'm already earning good starting pay. But when I get that license, I'll really be on my way." He doesn't have to wait three more years to make more money, though. "Every year we get an evaluation and a raise based on how well we did. Also, if I get an A in the apprenticeship program, I get a 4% raise—if I get a B, it's 3%, and so on," Kowalik explained. "That's a real good reason to study hard. Plus, the things I'm studying things I can immediately apply on the job."

Success STORIES

Kowalik has an advantage over the average student. "We use the NCCER CONTREN curriculum that I used in high school, so a lot of this year has been like a review, except there is a lot more added," Kowalik said. "For example, our field trip was to Eaton Electric where they make electrical service panels from scratch, and electronic control panels as well. After that trip, we had to go on the Internet and find articles about how the Eaton control panels are used and in what kinds of applications."

"I'M ALREADY EARNING GOOD STARTING PAY. BUT WHEN I GET THAT LICENSE, I'LL REALLY BE ON MY WAY."

A career start like Kowalik's is not unusual in the construction industry today. He took construction courses in high school and has managed to put that education to use immediately.

"Bill is impressive," said his foreman, George Booker. "I went through two other guys on this job before Bill came on. Those other guys didn't understand the concepts—they couldn't see how things go together. Bill's high school training gave him the big picture and a lot more. He has all the basic skills plus he knows how to work. Man, that's a blessing!"

In addition to the electrical training, Kowalik is getting to do some of what he thinks is "cool stuff." Booker has already taught Kowalik to operate a front-end loader, a forklift, and a trencher. "He can learn anything, and he learns to do things right," Booker said.

Kowalik's construction instructor at McEachern High, Kevin Ward, is not one bit surprised that Kowalik is doing so well. Ward sent Kowalik to apply at Wayne J. Griffin Electric. "Bill has a great work ethic and he's smart. He's book smart and worldly wise. He knows how to get along with people," Ward said. "Bill could have done anything he wanted to do: college, start his own business, anything. He's got a great future however he chooses to use what he's learning. I'm glad he chose the apprenticeship program, though, because he loves what he's doing."

"What's not to love about this? This is a great place to work. The people in the trades are super. Everybody on the site works together. If I need a 2-in. drill bit and don't have one, I can go borrow one from a plumber, or someone. The work is hands-on, everyday is a new challenge. And if I get dirty, nobody cares," Kowalik said. The tradespeople are here for a career. They're good people."

What's a workday like for Kowalik? It starts at 7 a.m. and ends at 3:30 p.m.. He goes to apprenticeship school one day a week for five hours. But when he's off, he's off. "I go home and get on the computer to check email and surf the

net some. I like to go bowling one night a week and my girlfriend and I play poker on Thursdays."

Where is Bill Kowalik's journey going to take him from here? "I am happy to be learning and earning at the same time. I know that if I drive by Floyd College some night, these light poles will be making the parking lot safe for the students and faculty. I'll know that I've done something that will last. Everything we do at Griffin is that way—done right and done to last. One day, I'd like to know enough to be able to run a whole job for my company. And someday, when I know enough about how things work, maybe I'll start Kowalik Electric." That has a nice ring to it, doesn't it?

Big dreams for a large man who is putting up giant light poles. But like those light poles that will light up a college parking lot, Bill Kowalik's plans and dreams are lighting the way to a bright future indeed.

Time to Build Your Career!
There are good-paying jobs and careers open to you in the construction industry. The following pages give more detail about these positions—the work, training required, pay, and career outlook.

Carpenter—cuts, fits, and joins wood and other building materials.

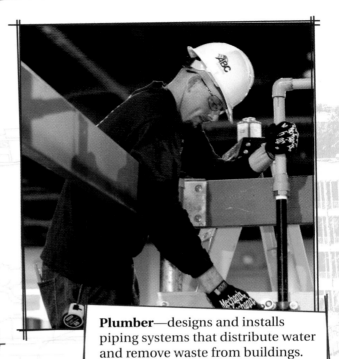

Plumber—designs and installs piping systems that distribute water and remove waste from buildings.

Mason—constructs walls of brick, concrete block, and stone.

Electrician—plans and installs wiring and electrical devices such as breaker boxes, switches, and light fixtures.

Welder—joins objects together by applying heat or pressure.

HVAC Technician—installs heating, ventilation, and air conditioning systems.

Heavy Equipment Operator—runs heavy machinery such as dump trucks, bulldozers, backhoes, front-end loaders, and excavators.

Electronic Systems Technician—designs, integrates, and installs products that carry voice, audio, and data signals.

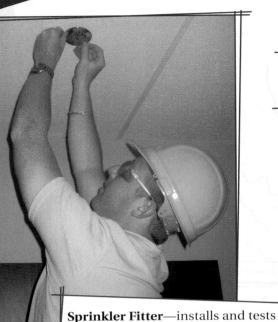

Painter—applies paint and coatings to structural surfaces.

Sprinkler Fitter—installs and tests automatic fire protection systems, including sprinklers, piping, and valves.

Sheet Metal Worker—cuts and molds sheets of metal into products for installing ventilation and air ducts, roofing, and gutters.

Ironworker—moves and places large girders, installs preconstructed ornamental materials, and strengthens concrete walls.

Mobile Crane Operator—hoists and swings loads with a rotating mobile crane.

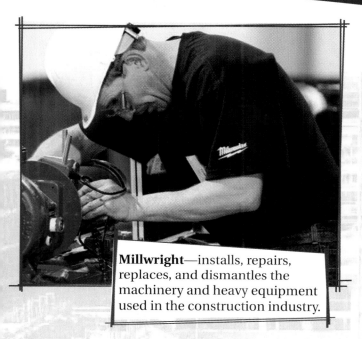

Millwright—installs, repairs, replaces, and dismantles the machinery and heavy equipment used in the construction industry.

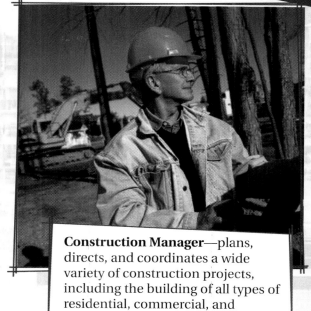

Construction Manager—plans, directs, and coordinates a wide variety of construction projects, including the building of all types of residential, commercial, and industrial structures including roads and bridges, wastewater treatment plants, and schools and hospitals.

Concrete Finisher—finishes horizontal surfaces such as airport and roadway pavements, and vertical surfaces including walls and cast columns.

Pipefitter—installs and repairs both high- and low-pressure pipe systems used in manufacturing, in generating electricity, and in heating and cooling buildings.

Carpentry is one of the world's oldest professions. Stone Age craftsmen used their axes to shape wood for shelters, animal traps, and dugout boats. By 3000 B.C., Egyptians were using copper tools to construct vaults, bed frames, and furniture.

Fast forward 5,000 years. Today's carpenters are well-trained, highly versatile members of construction's largest trade group. They must not only know about wood, but also how to work particleboard, gypsum wallboard, ceiling tiles, plastics, and laminates. They must master modern tools, fasteners, advanced construction techniques, and safety measures.

A skilled carpenter is never out of work. The career possibilities are unlimited.

THE WORK

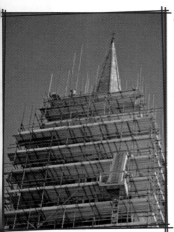

As a carpenter, you may be involved in many different types of job activities, from building bridges to installing kitchen cabinets. Depending on your employer and the type of project, you may specialize in one or two activities or be required to know how to perform many different tasks.

Your work will be active and sometimes strenuous. There may be periods of prolonged standing, squatting, kneeling, and climbing. You may work outdoors in adverse weather. And you will risk injury from slips and falls, contact with sharp or rough materials, and from sharp tools and power equipment. You must have a high regard for safety.

Here are the primary tasks of a carpenter:

☑ Check job and materials against local building codes

☑ Follow blueprints to lay out the job–measure, mark, and arrange materials

☑ Cut and shape wood and other materials with hand and power tools

☑ Join materials with nails, screws, staples, or adhesives

☑ Check accuracy of work with levels, rules, plumb bobs, or electronic devices.

TRAINING

Carpenters learn their trade through both formal and informal training programs. Formal apprentice programs combine on-the-job-training with classroom instruction, making it possible to earn while you learn. The traditional apprenticeship period is 4 years. You can get a jump on your training in high school by taking classes in English, algebra, geometry, mechanical drawing, and blueprint reading.

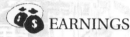
EARNINGS

The average hourly earnings of carpenters are $17, with those in the top 10% making more than $29 per hour.

CAREER OUTLOOK

Job opportunities for carpenters are excellent over the coming decade, particularly for those with the most skills. The need for carpenters is expected to grow as construction activity increases in response to demands for new housing, office and retail space, highway projects, and for modernizing and expanding schools and industrial plants.

> **"CARPENTRY HIT THE NAIL ON THE HEAD, SO TO SPEAK. I COULDN'T THINK OF A MORE FULFILLING CAREER CHOICE."**

Carpenters usually have greater opportunities than other trades to become general construction supervisors because carpenters are exposed to the entire construction process. And carpentry is an excellent path to starting your own business—about one-third of all carpenters are self-employed.

P Plumbing is another ancient trade. In 312 B.C., the Romans began bringing water to Rome by aqueducts, many of which still stand. Aqueducts were also used to carry waste for discharge into rivers downstream from the city.

Today's plumbers still have the job of providing systems to supply safe drinking water and remove wastes. As a plumber, you may also design and install systems to protect homes and business from fire, keep basements dry, bring gas to stoves, and extend municipal water and sewage lines.

 THE WORK

Plumbers install and repair water, waste disposal, drainage, and gas systems in homes and commercial and industrial buildings. They also install plumbing fixtures and appliances such as dishwashers and water heaters.

As a plumber you'll work with the latest technology, tools, and equipment. You'll need to keep up to date on technological developments, on the qualities and advantages of new materials, and on changes to the building codes and standards that govern all aspects of plumbing.

 TRAINING

Plumbing apprenticeships consist of 4 or 5 years of on-the-job training (earning a wage), plus 144 hours of classroom training per year. Classroom subjects include drafting and blueprint reading, math, applied physics and chemistry, safety, and plumbing codes and regulations.

Plumber

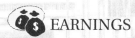 **EARNINGS**

According to a recent government survey, the average hourly wage for plumbers is $20. Those in the top 10% of the wage scale earn more than $34 an hour.

 **CAREER OUTLOOK**

Job opportunities are excellent for the plumbing trade. The demand for skilled plumbers, pipelayers, and steamfitters is expected to outpace the supply of workers trained in these crafts.

With experience, journeymen plumbers become master plumbers. Master plumbers are proven experts in the plumbing profession who are sought after to mentor less experienced craftspeople and to serve as team or project supervisors on the job. Skilled plumbers who also master small-business principles often go to work for themselves.

E Electricians are the ones who bring power to our homes and workplaces. As an electrician, you must be constantly aware of the hazards around you and make safety a habit.

 THE WORK

Electricians work with blueprints that indicate the locations of circuits, outlets, load centers, panel boards, and other equipment. During every step of a job you must follow the National Electrical Code® (NEC®) and comply with state and local regulations that apply to the system you are installing. Here are some of the things you'll be doing:

- ☑ Laying out the electrical system
- ☑ Obtaining needed materials
- ☑ Placing conduits (pipe or tubing) inside walls or partitions
- ☑ Fastening small boxes for switches and outlets
- ☑ Pulling insulated wires or cables through conduits
- ☑ Connecting wires to circuit breakers, transformers, outlets, or other components
- ☑ Testing for proper connections and electrical operation

Electricians work both indoors and out—at construction sites, and in homes, businesses, and factories. As an electrician, you must follow strict safety precautions at all times to avoid injury from electrical shock, falls, and cuts.

TRAINING

Most electricians learn their trade through apprenticeship programs, which usually last 4 years. Each year includes at least 144 hours of classroom instruction and 2,000 hours of on-the-job training (for which you'll be paid).

Training to become an electrician is offered by a number of public and private vocational and technical schools and training academies. Employers often hire students who complete these programs and usually start them off at a higher level than those without the training.

Experienced electricians are going back to school for classes on installing low-voltage voice, data, and video systems, which are becoming a part of everyday American life.

EARNINGS

A recent government survey lists the average wage for electricians as $20 an hour. Those in the top 10% earn more than $34 an hour.

CAREER OUTLOOK

The employment outlook for electricians is excellent for the coming decade. As the population and economy grow, more electricians will be needed to install and maintain electrical devices in homes, offices, and factories. Job opportunities will also be boosted by increased use of computers and telecommunications. And electrical skills are a good basis for starting your own company—about 1 in 10 electricians is self-employed.

M

Masonry is one of the world's oldest and most respected crafts. Archeologists have unearthed stone buildings dating back 15,000 years, and found bricks fashioned more than 10,000 years ago.

Today's masons have more than 100 different types of bricks, concrete block, tiles, and terra-cotta products with which to work. With skill and imagination, master masons are using these products to create structures that not only serve a functional purpose, but also stand as works of art—perhaps for future archeologists to marvel at.

 ## THE WORK

The term "mason" includes the crafts of brickmason, blockmason, and stonemason. Your work will vary in complexity from laying a simple masonry walkway to building an industrial furnace to installing an ornate exterior on a high-rise building.

Masons usually work outdoors. Work hazards include injuries from tools and falls from scaffolds, but these can be avoided by using proper safety equipment and procedures.

As a mason, you can expect to:

☑ Follow blueprints to lay out the job

☑ Master the use of trowels, special hammers and chisels, and diamond-bladed saws

☑ Understand the working properties of brick, stone, granite, tile, and other materials

☑ Cut or shape masonry materials to fit surface contours and openings

☑ Seal joints for a sealed, neat, and uniform appearance

☑ Use imagination to create unique surface designs and textures

Mason

TRAINING

Most masons pick up their skills informally by observing and learning from experienced workers. Many others receive training in vocational schools or from industry-based programs. Apprenticeship offers the most thorough training. Most programs require 3 years of on-the-job training, along with at least 144 hours a year of classroom instruction in subjects such as blueprint reading, math, layout work, and sketching.

EARNINGS

According to a recent government survey, the average wage for masons is $20 an hour, while masons in the top 10% earn more than $30 an hour.

CAREER OUTLOOK

Job opportunities for brickmasons, blockmasons, and stonemasons are very good over the next decade as the demand continues for new houses, industrial facilities, schools, offices, and other structures. Masons will also benefit from the increasing use of brick and stone for decorative work on building fronts and in lobbies and foyers. A career in masonry often leads to self-employment—about one-third of all masons have their own businesses.

How important is the Heating, Ventilating, and Air Conditioning (HVAC) trade? Nearly every building across the country has some form of equipment to providing heating and cooling, as well as air circulation and purification. And when one of these systems breaks down, no one is more in demand or welcome than an HVAC technician.

THE WORK

HVAC technicians install, maintain, and repair systems that control the temperature, humidity, and total air quality in residential, commercial, and industrial buildings. The work requires a thorough technical knowledge of these systems. Here are some of the tasks you'll perform as an HVAC technician:

As an HVAC technician, you'll work both indoors and out. You must take precautions to avoid electrical shock, and wear special safety equipment when handling refrigerants to avoid skin damage and prevent inhalation.

Here are some of the primary tasks you'll perform as an HVAC technician:

☑ Follow blueprints to install oil, gas, electric, and multiple-fuel HVAC systems

☑ Install fuel and water supply lines, air ducts, pumps, and thermostats

☑ Become skilled in using special tools, measurement gauges, and acetylene torches

☑ Connect electrical wiring and check units for proper operation

☑ Use combustion test equipment and analyzers to check for safe operation

☑ Recover and recycle chemical gases used in refrigerant systems

☑ Troubleshoot, diagnose, and correct problems throughout the entire system

HVAC Technician

TRAINING

Because of the increasing sophistication of HVAC systems, employers prefer to hire those with technical school training or those who have completed an apprenticeship. Formal apprenticeship programs normally last 3 to 5 years and combine paid on-the-job training with 144 hours of formal classroom instruction per year. A basic understanding of electronics is essential because of its increasing use in equipment controls.

EARNINGS

Recent government surveys show that HVAC technicians earn an average hourly wage of $17. Technicians in the top 10% earn $27 or more an hour.

CAREER OUTLOOK

Future job prospects are excellent for HVAC technicians, especially for those with training from an accredited technical school or a formal apprenticeship. Concern for the environment is driving the development for new energy-saving heating and air conditioning systems. This emphasis on better energy management will create more jobs as businesses and homeowners replace existing HVAC systems with newer, more efficient ones.

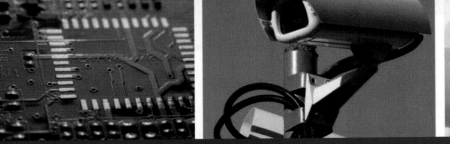

Construction's newest trade is that of electronic systems technician (EST). Consumer demand for electronic systems for security, improved communications networking and control, and automated lighting and energy devices is exploding. And this demand is being matched by the urgent need for skilled ESTs who understand and can install these advanced systems.

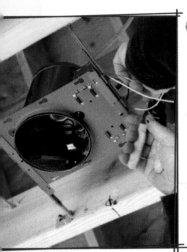

THE WORK

Electronic systems technicians install, connect, calibrate, and service products that carry voice, video, audio, and data inside homes, apartments, and commercial buildings. ESTs work for a variety of companies, including alarm, security, and home automation companies; custom designer/installers; data cabling companies; and home theater designers.

As an EST, you'll need a good understanding of electricity and electronics, microprocessors and computers, software programming, and signal/data communications.

The work of an EST is varied and challenging. Given the installation plan, you'll work independently without direct supervision. To avoid injury from electrical shock and other construction hazards, you must learn and follow strict safety precautions.

Here are the primary tasks of an electronic systems technician:

☑ Follow blueprints and construction specifications

☑ Identify wire pathways

☑ Pull, secure, and terminate wire and cables

☑ Install outlets and connection panels

☑ Install, test, and troubleshoot system electronic and mechanical components

☑ Program digital components to perform in the manner specified by the client

---

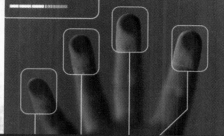

Electronic Systems Technician

TRAINING

ESTs receive the most complete training through formal apprenticeship programs, which typically last 4 years and combine on-the-job training with 144 hours of classroom instruction per year. In on-the-job training, you'll be under the supervision of a journeyman, working with others in real-life situations. In the classroom, you'll learn not only how things are done, but why.

EARNINGS

According to a recent government survey, the average hourly wage for an electronic systems technician is $17, while those in the top 10% earn upwards of $29 an hour.

CAREER OUTLOOK

The electronics systems industry is among the fastest growing in the world. All new construction includes electronic systems, and both residential consumers and businesses are replacing or modifying existing systems. The job prospects for electronic systems technicians have never been better, and the outlook is excellent for decades to come.

W

Welding is the most common way of joining metal parts. Heat is applied to the metal pieces, melting and fusing them in a permanent bond that is as strong as—sometimes stronger—than the metal itself.

Welding is used to join beams when constructing buildings, bridges, and other structures, and to join pipes in pipelines, power plants, and refineries.

 ## THE WORK

Welders have more than 100 different types of welding to use. The most common is arc welding, in which a powerful electrical circuit creates massive heat that fuses the metals and the steel core of a thin welding rod. As a welder, you will use similar techniques to perform soldering and brazing operations, which bond metals at lower temperatures.

As a welder, you will work both indoors and out, sometimes on a high platform or scaffold. You'll wear safety shoes, goggles, hoods with protective lenses, and other devices to prevent burns and eye injuries.

Here are the primary tasks of a welder:

- ☑ Plan your work from drawings and specifications
- ☑ Select the proper fusing materials, based on the metals involved
- ☑ Set up equipment and perform the planned welds
- ☑ Examine the welds to make sure they meet specifications
- ☑ Use the welding process as a cutting tool
- ☑ Become skilled in both manual and automated (robot-controlled) techniques

Welder

TRAINING

Training for welders can range from a few weeks of on-the-job training for low-skilled positions to apprenticeships involving several years of on-the-job training combined with formal classroom studies for highly skilled jobs. Formal training is available in high schools, vocational schools, and community colleges.

Prospective welders are encouraged to takes classes in blueprint reading, mechanical drawing, physics, chemistry, and metallurgy. Knowledge of computers is important for welding machine operators who are becoming responsible for programming their machines.

EARNINGS

Recent government surveys show that welders earn an average hourly wage of $15. Those in the top 10% earn $22 or more per hour.

CAREER OUTLOOK

Job opportunities for welders should be very good over the next decade as employers report difficulty finding enough qualified people. Technology is creating more uses for welding in the workplace. For example, new ways are being developed to bond dissimilar materials and nonmetallic materials, such as plastics, composites, and new alloys. Also, advances in laser beam and electron beam welding, new fluxes, and other new technologies and techniques all point to an increasing need for highly trained and skilled welders.

H

Heavy equipment operators are highly skilled workers who are needed on every construction site. They clear and grade land to prepare it for construction of roads, buildings, and neighborhoods. They dig trenches to lay sewer and other pipelines, and they hoist heavy construction materials. Heavy equipment workers also operate machinery to apply asphalt and concrete to roads and other structures.

 ## THE WORK

Heavy construction equipment includes bulldozers, forklifts, excavators, backhoes, graders, loaders, scrapers, and dump trucks, as well as many special-purpose machines, such as pavers, piledrivers, and tampers. Technologically advanced construction equipment has computerized controls and improved hydraulics and electronics, requiring more skill to operate.

Workers may choose to learn to operate a few of these machines or go on to become proficient in all of them. The more types of machines you master, the more likely you are to find steady employment and earn top dollar.

Heavy equipment operators work outdoors in every climate and weather condition. Operating heavy construction equipment can be dangerous, but you can avoid accidents by following proper operating procedures and safety practices.

Heavy Equipment Operator

TRAINING

Heavy equipment operators usually learn their skills on the job, starting with light equipment under skilled supervision, and then move on to operate heavy machines. However, formal training—in the form of an apprenticeship—provides more comprehensive skills on many more machines. Apprenticeship programs consist of at least 3 years of on-the-job training combined with 144 hours of classroom instruction per year.

EARNINGS

Government surveys show that construction equipment operators earn an average hourly wage of $17, and those in the top 10% earn upwards of $29 an hour.

CAREER OUTLOOK

It is difficult to imagine a construction project that does not need heavy equipment operators. Jobs for operators are expected to be plentiful over the next decade due to new construction, as well as the repair and replacement of the nation's roads and bridges.

Construction equipment operators often become instructors within companies or vocational schools. Also, operators make excellent managers for coordinating the availability of machines at the job site and supervising their servicing and repair.

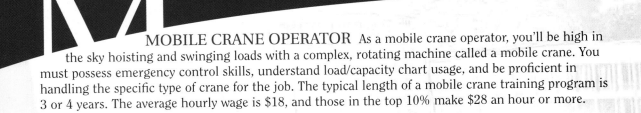

MOBILE CRANE OPERATOR

As a mobile crane operator, you'll be high in the sky hoisting and swinging loads with a complex, rotating machine called a mobile crane. You must possess emergency control skills, understand load/capacity chart usage, and be proficient in handling the specific type of crane for the job. The typical length of a mobile crane training program is 3 or 4 years. The average hourly wage is $18, and those in the top 10% make $28 an hour or more.

SHEET METAL WORKER

Sheet metal workers cut and mold metal sheets into products for installing and repairing ventilation and air ducts. You'll also be using your skills to construct and install aluminum siding, metal roofing, and gutters—maybe for your own home! The average length of a sheet metal training program is 4 years. You can look forward to an average hourly wage of $17, and if you're in the top 10%, you'll earn $30 or higher.

Other Trades

IRONWORKER

You have several choices as an ironworker. A structural ironworker moves and installs large girders and beams on bridges and building frames. Ornamental ironworkers install pre-constructed materials on elevators, stairways, and balconies. Reinforcing ironworkers strengthen the concrete in walls, piers, and roads.

As an ironworker, you must be skilled in math, blueprint reading, welding, and riveting. The average training period is 3 years, and the average wage is $20 an hour. Reach the top 10%, and you'll make $34 an hour or more.

PIPEFITTER

As a pipefitter, you will install and repair both high- and low-pressure pipe systems used in manufacturing, in the generation of electricity, and in the heating and cooling of buildings. You'll also install automatic controls that are increasingly being used to regulate these systems. Some pipefitters specialize in only one type of system.

You may enter the pipefitter profession in a variety of ways—through career and technical schools and community colleges plus on-the-job training, or through a formal 4- or 5-year apprenticeship program that combines on-the-job training with 144 hours of classroom training per year. The average wage is $20 an hour, reaching $34 for those in the top 10%.

SPRINKLER FITTER Sprinkler fitters install all types of fire protection systems, including the layout and installation of underground fire mains. As a sprinkler fitter, you'll follow blueprints to lay out and install hangers and overhead piping in all types of buildings, including high-rises, warehouses, aircraft hangers, hotels, motels, and homes.

The sprinkler fitter apprentice program is 4 years long, requiring 8,000 hours of on-the-job training over 4 years, plus 144 hours of classroom training per year. A journey sprinkler fitter earns $20 an hour, while those in the top 10% in the profession earn $34 an hour. Wages will vary among different geographical areas of the country.

CONCRETE FINISHER

This craft will find you building skyscrapers, sidewalks, houses, highways, and dams. You'll put the finishing touches on horizontal surfaces such as airport and roadway pavements, floors, and foundations. You'll finish vertical surfaces such as walls, cast columns, piers, and girders. Good math skills will be needed to calculate vertical rise or fall, squareness, and right triangles. The average concrete finishing program is 2 or 3 years long. The average wage is $16 an hour, while those in the top 10% make upwards of $26 an hour.

PAINTER

Painters are part of every construction team—applying paint and coatings to structural surfaces. They must know what type of paint or coating to apply and how the environment will affect the surface. You'll work with a variety of hand and power tools to prepare the surface and apply the paint or coating. The average length of a painting training program is 3 years. As a painter, you'll earn an average hourly wage of $14, and make more than $23 if you're in the top 10%.

MILLWRIGHT

Millwrights install, repair, replace, and dismantle complex machinery and heavy equipment used in many industries and on construction projects. The wide range of facilities and the development of new technologies will require you to continually update your skills as a millwright—from blueprint reading and pouring concrete platforms for machinery to diagnosing and solving mechanical problems. Typical training programs are 4 to 5 years. Millwrights earn an average hourly wage of $21, and those in the top 10% earn upwards of $32 an hour.

CONSTRUCTION MANAGER

Construction managers plan, direct, and coordinate a wide variety of construction projects, including the building of all types of residential, commercial, and industrial structures, roads, bridges, wastewater treatment plants, and schools and hospitals. As a construction manager, you may oversee an entire project or just part of a project. You probably won't play a direct role in the actual construction of a structure, but you'll schedule and coordinate all design and construction processes, including the selection, hiring, and oversight of specialty trade contractors.

If you're interested in becoming a construction manager, you'll need a solid background in building sciences, business and management, as well as extensive work experience within the construction industry. Managers earn an average yearly salary of $70,000, and those in the top 10% make upwards of $125,000.

Success STORIES

SPRINKLER FITTING A GOOD FIT

Elizabeth Kelly proudly explains she's a third-generation sprinkler fitter.

"My grandpa was a sprinkler fitter in Chicago. My dad is a sprinkler fitter. So that makes me a third-generation sprinkler fitter," she said.

So it's not a big surprise she decided to follow in her family's footsteps. What may be a surprise is her certainty. With a rare focus, she explained her passion for the trade. "I always knew this is what I wanted to do," Kelly said. "I knew when I finished high school this is what I wanted to do."

After she graduated from Berkmar High School in Gwinnett County, Georgia, in May 2004, 18-year-old Elizabeth found the CEFGA/GFSA Sprinkler Fitting Apprenticeship Program on the Internet. She enrolled and today is a Level 1 Sprinkler Fitting apprentice. "The classes are great. Our instructor explains everything. He's patient. I feel like I am learning so much about the sprinkler industry," she said. "And whatever I don't get in class, I can always go home and ask my dad."

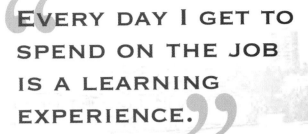

> "EVERY DAY I GET TO SPEND ON THE JOB IS A LEARNING EXPERIENCE."

Her dad is Jerry Kelly, who works with Property Engineering Services of Marietta, Georgia.

Kelly said she has learned a lot about the sprinkler industry just by being around her dad. "I always begged my dad to let me work with him on the weekends. Over the summer, I changed out pipe, installed heads," she said. "Every day I get to spend on the job is a learning experience."

Women make up fewer than 5% of the construction trades but that doesn't seem to matter to Elizabeth. "I kind of like walking out onto a construction site and being the only girl. Sometimes they (the other sprinkler fitters) laugh at me, because I don't know how to do everything. But, I'm learning, and they usually offer to help."

FAST TRACK FOR CAREER IN MASONRY

Shaun Fitzgerald drives race cars. Almost every weekend you'll find him at the Senoia, Lanier, or Atlanta Motor Speedway. He drove his first race in a Quarter Midget at age four and a half. These days he has two race cars, a 1934 Ford coupe and a 1937 Ford sedan, both of which he races in the Pro Division—Legends Category. Not bad for an 18-year-old, but being a race car driver is not Fitzgerald's only ambition.

On this day, he is loading his chalk block in preparation for his masonry class which is about to begin at the Masonry Association of Georgia's Apprenticeship Program in Avondale, Georgia. Race cars and masonry? What's the story here? "When I was a senior at McEachern High School,"

recalled Fitzgerald, "the Apprenticeship Coordinator here at the Masonry Association School, came in one day to talk to us about careers in the masonry field. It sounded pretty good to me, and it's turned out even better than I expected."

Fitzgerald got a job with Cornerstone Masonry. He works at the Atlanta airport on the new runway and concourse renovation projects. "I started out as a laborer, like everybody else, but every chance I get, I get 'on the wall' and lay some block," said Fitzgerald. "My foreman lets me do it because he knows I'm in the masonry apprenticeship program. He stays with me and makes sure I'm doing things the right way. Most of the time I do whatever needs to be done, but I'm learning a lot every day." The term "on the wall" means actually getting up on a scaffold with the masons and laying concrete block that will be there for at least a couple of hundred years. "Yeah, that's pretty neat knowing you're doing something that is going to last a lot longer than you'll even live," Fitzgerald said with a satisfied smile.

Wait a minute here. A laborer who's just starting in a masonry training program and is doing production work on something as important as the new runway? Because Fitzgerald completed Masonry Levels 1 through 3 in his high school construction class, he can do the work. The high school used the National Center for Construction Education and Research (NCCER) curriculum, and the instructor submitted the training information to a national

"ONE OF THE MOST IMPORTANT THINGS I LEARNED IS HOW TO WORK SAFELY ON THE JOB."

database so that any employer can determine an apprentice's training.

Kevin Ward, the construction teacher at McEachern High School in Powder Springs, Georgia, has been certified to teach the NCCER curriculum by the Construction Education Foundation of Georgia (CEFGA), a sponsoring agency of the NCCER. "You have to attend an intensive two-day workshop and pass a practical exam," Ward explained. "Then you have to stay current with CEFGA and the NCCER by submitting training module completions to the NCCER database. It is a pretty rigorous process, but it is important to do this for our students because that's how an employer knows the training has been done correctly. In Shaun's case, I really didn't have to do much training, per se. I taught Shaun how to learn and let him go at it. He can teach himself anything. He's smart, honest and hard working. Plus, and this is a big plus, Shaun has a flair for leadership."

"I learned all the basic construction skills in Mr. Ward's classes, but I got to specialize in masonry," Fitzgerald said. "One of the most important things I learned is how to work safely on the job. Everyone has to be aware of safety and work responsibly if everyone is going to be safe on the job."

Success STORIES

In addition to the basics and safety, Fitzgerald also learned about good work ethics. Kevin Ward notes, "We work directly with industry to make sure we're teaching the skills employers are looking for. We have a council of industry leaders who act as an advisory board to the construction program. Industry people come into the classroom and talk to the students about the real world of construction. Every single speaker tells us about the importance of a strong work ethic. That makes an impression."

Shaun Fitzgerald didn't just take the education and run, either. He goes back to McEachern frequently to tell the construction students about how it is in the real world.

He also works with the masonry students to demonstrate all the new tips and techniques he is learning in the apprenticeship program and on the job.

> " IN THE APPRENTICESHIP PROGRAM, I'M POLISHING THE SKILLS I HAVE AND LEARNING NEW ONES. "

When asked where he wants to go with his masonry career, Fitzgerald said, "I want to move up through the ranks. I can see myself in the foreman position within a few years. I'm not saying it will be easy, but I know what I have to learn. In the apprenticeship program, I'm polishing the skills I have and learning new ones. But to become the foreman, I'll have to learn estimating, labor management, planning, how to work closely with the general contractors, and more. Cornerstone Masonry is a good place to learn all these skills. The people are friendly and want to help you succeed."

Mark Bryson, co-owner of Cornerstone Masonry, is glad Fitzgerald came on board. "He's a good kid. He's good natured. He works hard. He shows up and he doesn't play around. He's just an all 'round good guy," Bryson said. "We'd like to get him laying block full time, and one day we'd like to see him running a job."

But there are other ambitions. "I live near a neighborhood where the homes have a lot of stonework in the European style architecture," Fitzgerald said. "Not just stones slapped up for curb appeal, but real works of art. I drive through there often and look at the artistry of the design and workmanship. I am going to learn to be a stone mason, too. I want to do work like that. It's beautiful, and it will last."

Is anyone surprised that this race car driving mason has an artistic flair? "Not in the least," said Ward. "Shaun decided he wanted to construct an archway as a project in the Level 3 NCCER training. No one had done anything like that before. Shaun figured out how to make the form for the concrete arch, poured concrete into the form, and set brick on it. It was a work of art itself!"

Whether Shaun Fitzgerald becomes known as a race car driving artistic stone mason or a stone mason artist race car driver doesn't really matter. What matters is that this young man knows where he's going. And, he knows what he needs to know to get there.

How Do I Get Started?
You can start building a rewarding, well-paid career in construction this very day! Whatever your educational background, there are people and programs ready to help you on your path to success. Let's look at some of the options.

HIGH SCHOOL PROGRAMS

HIGH SCHOOL PROGRAMS Tomorrow's jobs—including construction—will require more knowledge, better skills, and more flexible workers than ever before. Training for these jobs begins in high school with basic courses in English, algebra, trigonometry, chemistry, and physics.

Many schools offer electives in construction technology, drafting, and design. And in some communities, schools have established vocational-technology centers offering instruction and in-the-field practice in carpentry, masonry, plumbing, and HVAC technology.

Ask your career and technical educator or occupational specialist to help design a program to meet your goals. And while you're there, check out opportunities for job-shadowing, internships, and mentoring programs in the construction industry.

ON-THE-JOB TRAINING

Many workers with few skills enter the construction industry by obtaining a job with a contractor who will then provide on-the-job training. Entry-level workers generally start as laborers or helpers, assisting more experienced workers.

Beginning workers perform routine tasks, such as cleaning and preparing the work site and unloading materials. When the opportunity arises, they learn from experienced construction trades workers how to do more difficult tasks, such as operating tools and equipment. During this time, the construction worker may elect to attend a trade or vocational school, or community college to receive further trade-related training. They may also take steps to enter an apprenticeship in a registered program.

APPRENTICESHIP

Apprenticeship is the formal path to earning certification as a trades journeyman or journeywoman. Apprenticeships are offered to qualified individuals by employers, employer associations, and joint labor-management organizations.

Apprenticeship combines on-the-job training with related technical instruction. Registered programs have term lengths ranging from 1 to 5 years, although 3 or 4 years is the most common requirement.

How Do I Get Started?

Apprentices are expected to complete 2,000 hours of supervised on-the-job training and at least 144 hours of in-class instruction per year.

As an apprentice, you'll be paid as you work. Wages usually start at 50% of the journey-level wage and gradually rise as your skills increase—eventually reaching 90% during your final training period. You'll also:

- Have full-time employment
- Learn a trade with skills that will serve you a lifetime
- Achieve the higher wages and increased career opportunities that come with journey status
- Be employable anywhere

COMMUNITY COLLEGES

Community colleges put higher education close to where you work and live. They have an open admissions policy and offer many courses, including ones that will help you prepare for a bright future in the construction industry.

Community colleges are flexible. Nearly half of their students work full time, so they offer courses at convenient times. When you combine these factors with a low tuition rate, community colleges are an excellent place to gain construction job skills.

More and more, construction apprenticeship programs are linked to community colleges, most of which offer two-year, associate-degree programs in construction, construction science, construction technology, and construction management.

CAREER COLLEGES

Career colleges offer another path to gaining the skills needed for success in the construction industry. These for-profit institutions provide professional, career-specific educational programs in hundreds of occupational fields, including construction.

Career colleges graduate approximately one-half of the technically trained workers who enter the U.S. workforce. Depending on your course of study, you can earn anything from a short-term certificate to a full doctorate degree. To find career colleges in your area, search the web for "Career Colleges." You'll get a listing of websites that let you explore local career colleges and the construction industry-related programs they offer.

FOUR-YEAR COLLEGES If you decide to go to regular, 4-year college, there are many programs that offer undergraduate degrees in construction-related fields. Managerial personnel in the construction industry usually have a college degree or considerable experience in their specialty.

Four-year colleges also offer courses such as building science, business and management, contracts, scheduling, and construction methods, materials, and regulations needed by managers to build a solid background for success in the construction industry.

The highest level of construction education is the graduate degree. A master's degree opens the doors to a career in executive management.

CONSTRUCTION CAREER PATH

Once you get started, the construction industry will offer you many career choices and opportunities—craft professional, foreman, project supervisor, project manager, company owner, just to name a few.

Take the first step today and join the more than 7 million men and women in construction who are building not only rewarding careers, but also lasting monuments to their skill and dedication.

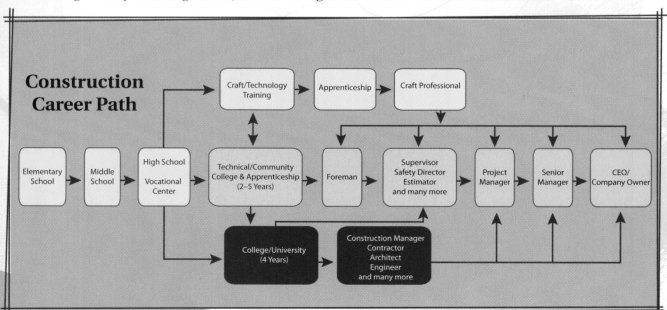

A REMOTE LABORATORY

For Paul Jeffries' high school HVAC students, they have the best of both worlds: a top-notch classroom education where they learn the theory behind proper installation of heating, cooling, and ventilation systems, and the opportunity to design, fabricate, and install an HVAC system into a real house.

All students enrolled in the construction trade courses at **Thomas Edison High School of Technology** in Silver Spring, Maryland, get the ultimate hands-on experience of building a house, from designing and drafting to pounding the last nail, connecting the last wire, and laying the last brick.

Edison has held construction trades classes for more than 20 years, said Linda Lynch, School and Community Outreach Resource Teacher. The curriculum includes a semester-long introduction course, and two-year programs in carpentry, HVAC, masonry, electrical, and plumbing. Typically tenth-graders enroll in the introduction course and then choose a two-year program to enroll in as juniors and seniors.

Each year, the students design and build a home with the support of the county's Construction Trades Foundation, which acquires the land and stands behind the finished product. The Foundation, founded 30 years ago, includes professionals in the trades—from architects, who choose one of the students' designs, to contractors who oversee the work on the house throughout the year. The plumbing students install the plumbing, the masonry students lay the bricks, and the HVAC students design and fabricate the HVAC system—ducts, refrigerant lines, controls equipment, and everything in between. "My students spend 25 to 30 days working on the house project, while the carpentry students spend 120 days of 185 school days on site," Jeffries said. "I suppose about 75% of the house is built by the students, the rest by professional tradespeople."

In 2006 the students will have built their 35th house. Last year's sold for just under $600,000. Before the house is sold, an open house is held for students, instructors, community leaders, and citizens. Also, a cornerstone is placed at the site dedicating the home.

"The quality is second to none," Jeffries said. "If any part of an installation is in any way substandard, we have the luxury of time and their 'free' labor to do it and re-do it until we get it just right. It's a great learning experience."

Both eleventh- and twelfth-graders work on site, and in some cases, the seniors are able to mentor and manage the younger students, getting valuable project management experience. "It's a great lab for kids because they have to learn how to work with each other and the other trades, and sometimes the experience is sufficiently 'real life' to allow the students to decide whether this kind of work is a valid career path. They don't have to guess—they know from experience," said Jeffries.

Students must apply for admission to Edison, and acceptance is based on their grades and their attendance record because of the hands-on nature of the curriculum. "You can't make up this kind of work," Lynch said.

Success STORIES

Once accepted, students attend three class periods at Edison and the rest of the day at their co-enrolled high school. "What's good is that it doesn't cut them off from their co-enrolled school," said Lynch. "Students can still participate in sports, music, dance, and other extracurricular activities." Students who attend Edison graduate from their co-enrolled high school just like any other high school student in the county, but they have the huge advantage of having gained a marketable skill, Lynch said.

"The response from students has been tremendous," Jeffries said. "As in any school, some kids decide it's not for them, or have trouble completing the program. But most excel. HVAC employers come to my classroom, talk to my students, and conduct job interviews even before they've graduated," Jeffries said. "Many of the kids work after school for contractors at as much as $11 or $12 an hour. By the time they are graduated, they are advanced entry-level employees," Jeffries said. About one-third go on to college to get degrees in engineering or other areas, but most plan to go straight to the workforce from Edison.

"And the workforce needs it," said Steve Boden, Executive Vice President of the Construction

Trades Foundation, which oversees not only Edison's program, but technical education for all 140,000 high school students in Montgomery County outside Washington, DC. "There is a huge demand from the industry for people with skills like these. The house is a great way for students to show prospective employers what they can do."

Going to college and getting an education is important, Lynch said, but there is a place for this. "It's important to go as far as you can in your education. But that's not the only way," she said. "These students get hands-on experience and acquire skills so that when they graduate, while other kids may be getting a job at a fast food restaurant, our kids have real skills."

CONSTRUCTING A LIFE

Cal Pygott likes to think of his high school shop class as more than just teaching his students the rules of construction. It's more about teaching them the rules of life.

"When you walk on to that job site, you are officially an adult," the 16-year teaching veteran said he tells his students. "You'll be working with 30-, 40-, and 50-year-olds that will only treat you like an adult if you act like one."

Pygott has been instrumental in developing the Construction Academy at Bothell High School in the Northshore School District outside Seattle, Washington. He has implemented NCCER's curriculum at the school, and to date, Bothell High School is the only school in the state that offers full NCCER certification.

"I run the program differently than anybody else," Pygott said. After 24 weeks of classroom learning, students transition to paid internships on construction sites where they get hands-on experience. More than 75% of students stay on permanently with the contractors with whom they are placed. "Students have been through the book-learning, but they are still a little green on the construction principles and this gives them a chance to get practical experience."

But really, the carpentry is the easy part, Pygott said. The real challenge has nothing to do with carpentry. "Kids don't know anything about being an adult. All they see is that they get to spend money and drive their own cars. So we practice being adults."

You need to show up on time for work. So to prepare his students, Pygott locks his classroom door once the bell rings. "If you're late, you're going to have to explain first because that's what you'll need to do at work." When students work in shop class, they wear hard hats, safety goggles, and any other protective equipment that is necessary—just as they would on a job site. "And no inappropriate clothes, like jeans you're going to be tripping over, or t-shirts that have some offensive message on them."

Pygott explains the necessary tools students should bring with them to their first day: driver's license and social security card, for example. "Kids don't know these things—we need to practice."

And while there will always be students who struggle with completing their internships, the vast majority of students excel. When the program was implemented at Bothell 7 years ago, 10 or 15 students signed up for the 24 spots in the class. Today, Pygott turns away students each year because the course is so popular. And he said it hasn't been a challenge to convince students that construction is a worthy career path, but rather, their parents are the tough sell.

"There is a stereotype from the '50s and '60s that people who go into construction do so because they can't do anything else—they are uneducated and have no options," Pygott said. But actually the opposite is true.

The average age of a construction worker currently is 49, and the market is in need of younger workers, Pygott said. In the next 6 years, the demand for skilled construction workers will reach one million.

Parents also often believe their kids will struggle financially in the construction trades. Not so, Pygott said. At $25 to $30 an hour, a worker can provide for his or her family. Not to mention the unlimited options available in advancement. "Just because you started out banging a hammer doesn't mean you'll always have to if there's a different way you want to go," he said. "You don't need a college degree to be a project manager or a superintendent."

About one-third of Pygott's students do go on to college to seek degrees in engineering, industrial hygiene, or architecture. "But if you go on to architecture, you'll know how a building goes up so you'll have a leg up." More importantly, "it's an industry where you can move just about anywhere and find a job, move in any direction you want, and you can look from the side of the road at a building you help put up, and say, 'I did that' and take pride in what you've done."

> **THE AVERAGE AGE OF A CONSTRUCTION WORKER CURRENTLY IS 49, AND THE MARKET IS IN NEED OF YOUNGER WORKERS...**

www.nccer.org

If you're seriously considering a career in the construction industry, the first place to look for information and opportunities is in your own community.

Local Sources. You'll find a network of individuals, organizations, and commercial firms ready to help you get started in a program or—possibly—your first construction job. Be sure to check out:

- friends and family—let people know you're interested in construction, they'll help
- school officials and career advisors—it's their job to help you start your career
- Chamber of Commerce—they have their finger on local construction activities
- construction companies—ask, and you might get a walk-on job!
- builder associations—their goal is to get skilled workers—you—into the trades
- vendors—places like Home Depot and Lowe's may have some job leads—ask!

Internet Sources. The internet—or the web—is a valuable source of information. For example, if you search Google for the topic "Careers in Construction," you will instantly get more than 80 million listings! If you don't have a computer, ask your school's career and technical educator or school librarian, or go to a public library, for help in your search. You can find almost anything about the construction industry on the web, including:

- complete and current information about each trade
- construction career options and pay ranges
- industry growth forecasts
- who is hiring construction workers, including companies near where you live
- apprenticeship opportunities
- local community colleges and the construction courses and degrees they offer

The following websites will be especially helpful as you begin your search for the best path to a rewarding career in construction.

www.nccer.org—this is the website for the National Center for Construction Education and Research, a not-for-profit education foundation established in 1995 to help build a safe, productive, and sustainable workforce of craft professionals.

NCCER develops and publishes the Contren® Learning Series, consisting of construction, maintenance, and pipeline curricula; safety programs; management education; and skills assessments. The series is taught nationwide by accredited NCCER sponsors, including contractors, owners, national trade associations, government entities, and for-profit schools. In addition, thousands of secondary and post-secondary schools teach with Contren®.

Once you're at the NCCER site, select School-to-Career, and then click Links. This will bring up a valuable list of other websites you can go to for more information.

nccer.monster.com—NCCER has partnered with Monster®—the web's premier clearinghouse for job information—to provide a one-stop career resource center for the construction industry. In addition to a resume builder and job search function, this career site includes up-to-date information on construction wages, top construction crafts, an industry career path, and useful links for instructors and students.

www.SkillsUSA.org—SkillsUSA is a national nonprofit organization serving more than 280,000 high school and college students and professional members enrolled in training programs in trade, technical, and skilled service occupations.

Explore the SkillsUSA website for information on competing in national skills championships, and other programs to:

- increase student awareness of quality job practices, attitude, and employability skills
- increase opportunities for employer contact and eventual employment
- ensure that students enrolled in technical education are training in the skills needed by employers.

Wwww.ABC.org—Associated Builders and Contractors is a national association representing 23,000 merit shop construction and construction-related firms in 79 chapters across the United States. ABC's membership represents all specialties within the U.S. construction industry and is comprised primarily of firms that perform work in the industrial and commercial sectors of the industry. Local ABC chapters and members offer craft training as well as registered apprenticeship programs in a variety of skilled trades.

On the ABC home page, select Training & Education, and then click Try Tools to learn about career opportunities in construction from the skilled trades to design and construction management. There are separate sections for younger students, high school students, and teachers and parents.

www.AGC.org—The Associated General Contractors of America is the nation's largest and oldest construction trade association. AGC is dedicated to improving the construction industry by educating the industry to employ highly skilled workers, promote use of technology, and provide quality projects for owners—both public and private.

On the AGC home page, select Other Industry Issues, and then click Education and Training for a wealth of information on training programs, student competitions, and other opportunities for advancing your career in construction.

www.careervoyages.gov—This site is sponsored by the U.S. Department of Labor and U.S. Department of Education. It is designed to provide information on high-growth, high-demand occupations, along with the skills and education needed to get those jobs.

From the home page, select Construction for a wide range of information on getting your career started. This site, which can also be seen in Spanish, lets you search for apprenticeship opportunities, and community colleges and 4-year colleges that offer construction programs in your zip code or state.

WELL ON HER WAY TO BUILDING SOMETHING

Jennifer Smith's mom was both amused and a little concerned. Amused because 5-year-old Jennifer had taken apart every toy she owned to see how it worked and put them back together. But Mom was a bit concerned because Jennifer had announced that she was going into the construction business when she grew up. That's pretty nontraditional for a child, especially a girl. "I'm going to build houses," little Jennifer declared with authority. "She'll probably outgrow it," Mom thought.

Jennifer hasn't outgrown it. In fact, she has grown into it. Jennifer is a sophomore at Florida A&M, majoring in Construction Engineering Technology. She knew where she was going at age five, and now she's well on her way.

A significant step in her journey was the four years of construction classes she took in high school. "Jennifer's only question was about which discipline she wanted to specialize in," said Kevin Ward, a teacher at McEachern High School in Powder Springs, Georgia. "She was absolutely certain that construction was her career choice. I don't ever recall any other freshman student who knew so clearly what she or he wanted to do in life."

Jennifer breezed through the core curriculum of safety, hand tools, power tools, construction math, blueprint reading, and rigging. Then she looked at masonry, carpentry, and plumbing, but chose electrical as her field of study.

Why electrical? "Problem solving," Jennifer said without hesitation. "Say someone wants switches on three different walls to control a light and fan in the middle of a room. I love to figure out the easiest way to lay out the

Success Stories

> ## "THERE'S ONLY ONE BEST SOLUTION, AND I LOVE TO FIND IT."

circuit and use the least amount of wire. You have to factor in what else is going to be on the circuit and how the source wiring run is going to be made from the breaker box. There's only one best solution, and I love to find it."

Success STORIES

"Jennifer's interpersonal skills are most amazing," Ward said. "I've never seen anyone so focused. She was awarded a Dual Seal Diploma from McEachern. She was inducted into both the National Honor Society and the National Technical Honor Society. She was instrumental in starting the National Technical Honor Society chapter at our school and she was active in the local chapter of both honor societies."

Jennifer's electrical interests expanded into industrial motor controls. She won first place in the Georgia SkillsUSA industrial motor controls contest twice.

"You just can't imagine the thrill of being the lone woman competing against the best men in the state and winning first place," she said. "I also got to compete in the National SkillsUSA contests in Kansas City. I'll never forget the excitement of those three trips," she said. Three trips? "Yeah. I came in second in the state contest my sophomore year, so I got to go that year, too." In her senior year, Jennifer placed 12th in the nation.

Now that she is in college, Jennifer's focus has narrowed a bit. She has decided she wants to pursue construction management.

"I was President of my SkillsUSA chapter for two years. I found out you can get a lot more done by organizing a group of people toward a common goal than you can get done by yourself. That's why I chose Construction Engineering Technology as my major.

I am learning the tools I need to organize projects and teams of people. I love to work with a group of dedicated people to get a job done, and done right," she explained. "I'm really looking forward to my internship this summer so I can practice using some of the skills I've learned."

Asked how she found out how to get an internship, Jennifer exuberantly described the placement programs at Florida A&M. "The school works hard to help us find jobs. A lot of companies come onto campus and talk to us about what's available. That's how I found Hensel Phelps Construction Company, with whom I'll be doing my summer internship. I don't know where I'll be this summer. They have jobs all over the country. But wherever I am, I'm going to be building something and using the skills I've learned here at school. Wow! What an adventure!"

Some students are going to be lying around this summer wondering what they're going to do with their life. Jennifer Smith is going to be doing with her life what she's wanted to do since she was five—construction.

ON HIS WAY UP

Jared Garber likes watching the lights come on. The 20-year-old electrician is part technician, part project manager, and part problem-solver. His favorite part of a big job is in the end, flipping the switch, and seeing the room light up. "Some days I really need to think a lot and problem-solve," Garber said. "But it's great to see everything come together in the end."

Garber doesn't really recall what drove him into the electrical trade. He just remembers that as an eighth-grader, he one day decided to shadow a cousin who was an electrician. "I just really found what he did interesting, so I called him and set it up," he said. That led him to enroll in the Miami Valley Career Technology Center in Clayton, Ohio, his junior and senior years in high school.

The accolades came quickly for Garber, who showed great talent at an early age. As a 16-year-old high school junior, he competed in his first national SkillsUSA Championship where he placed first in his craft. "It was a mini-project that included several tasks of things I would have to complete in the field," he said. "That's what really got the ball rolling for me," leading Garber to work for a small electrical contractor while in school.

" I ACTUALLY MAKE MORE MONEY THAN MANY OF MY FRIENDS WHO HAVE COLLEGE DEGREES. "

Upon graduation, Garber began working for Beacon Electric Contractors in Cincinnati, and is expected to complete the ABC Apprenticeship program and receive his journeyman's certification in 2006. Earlier this year, Garber competed in the ABC National Craft Championship where he placed first in his trade again.

The 20-year-old balks at the contention that young adults can't make money in the construction trades. "I would totally disagree with that," he said. "I actually make more money than many of my friends who have college degrees." But Garber hasn't ruled out college. In fact, he has earned some credit towards a construction management degree, and hopes to get his degree one day. "I just have a lot on my plate," he said. He recently bought a home and is engaged to be married. For now, he enjoys working his way up the ladder at Beacon and helping his parents on their 1,500-acre farm.

"How well you apply yourself and hard you want to work determines how far you will go," he said.

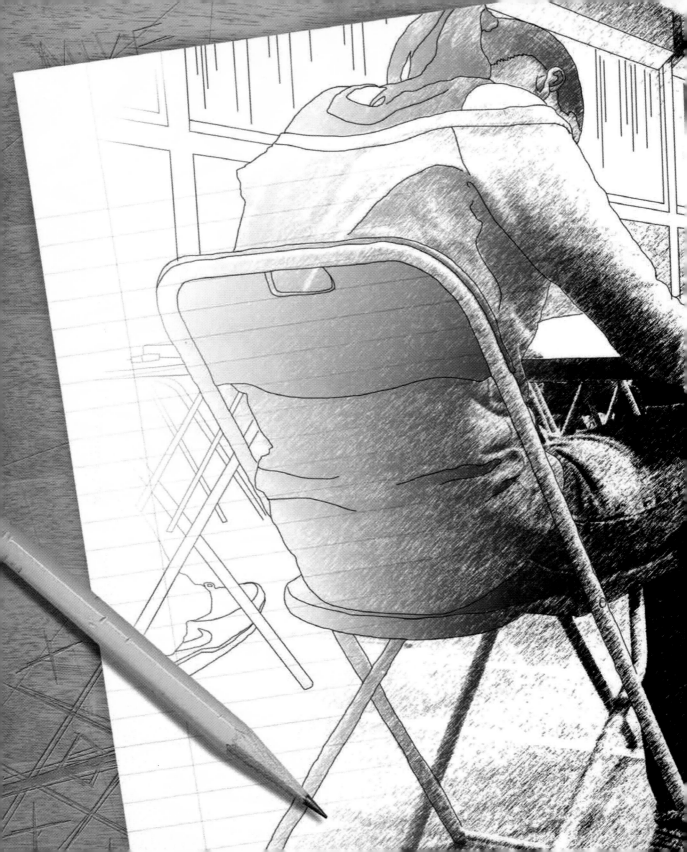

COME JOIN THE CONSTRUCTION TEAM!

The Bureau of Labor Statistics (BLS) says we'll need 250,000 new construction workers this year and each year after that to meet the demand for new homes, schools, office buildings, nursing homes, industrial plants, and high-tech medical facilities.

The BLS has more good news. It reports that construction workers earn more than workers in other industries. Construction workers have an average hourly wage of $18.51 as compared to $15.03 for all workers in the private sector, and $17.75 for all occupations.

START YOUR CONSTRUCTION CAREER TODAY

There's a construction job waiting for you today. You just have to decide to go for it and then take the first step. And whether you start as a laborer and work your way up, or become an apprentice and gain your skills through classroom instruction and on-the-job training with pay, you'll join a team of 7 million men and women who are proudly building the America of tomorrow.